© Copyright 2025 Stony Brook.

All rights reserved. No part of this publication may be reproduced, stored in a retrieval system, or transmitted, in any form or by any means, electronic, mechanical, photocopying, recording, or otherwise, without the written prior permission of the author.

Order this book online at www.trafford.com
or email orders@trafford.com

Most Trafford titles are also available at major online book retailers.

 www.trafford.com
North America & international
toll-free: 844 688 6899 (USA & Canada)
fax: 812 355 4082

Our mission is to efficiently provide the world's finest, most comprehensive book publishing service, enabling every author to experience success. To find out how to publish your book, your way, and have it available worldwide, visit us online at www.trafford.com

Because of the dynamic nature of the Internet, any web addresses or links contained in this book may have changed since publication and may no longer be valid. The views expressed in this work are solely those of the author and do not necessarily reflect the views of the publisher, and the publisher hereby disclaims any responsibility for them.

Any people depicted in stock imagery provided by Getty Images are models,
and such images are being used for illustrative purposes only.
Certain stock imagery © Getty Images.

ISBN: 978-1-6987-2033-3 (sc)
ISBN: 978-1-4907-9036-7 (e)

Library of Congress Control Number: 2018909608

Print information available on the last page.

Trafford rev. 10/13/2025

IN COMMON

A Unified Theory

OF

Every Thing

— **Stony Brook**

A Grand Unification Theory

of

Gravity, Relativity, and Uncertainty

"I want to know God's thoughts….the rest are details."

—Albert Einstein

What If ... .   .

1. All Mass, springs from the fabric of space-time itself.

2. Gravits, Tonne, and Uncertainty, are the three most fundamental and indivisible entities of Existence.

3. Gravits, are the grains of space. They are the fundamental individual units of Mass, as well as the smallest, and the densest. Gravits are likely to take the shape of incompressible spheres.

    All Gravit comprise approximately 75% of the fabric of space-time.

4. Tonne, are the gaps of null space that lie between each Gravit. They are the fundamental units of Time. They make it possible to measure movement, and are the cause of its fluidity.

    All Tonne, comprise approximately 25% of the fabric of space-time.

5. GraviTime, is synonymous with space-time. It is the stuff of which all things in the Universe are made.

    All Mass and Force, are merely questions of the quantities of GraviTime and the movement of its components—Gravit and Tonne. Together, they form tightly packed spheres, along with the gaps of null space between them.

6. Normal GraviTime, is the fabric of the space-time vacuum itself. All objects and matter particles, are constructed of clustered knots of cohered movement, which are made from this fabric.

    Matter GraviTime, are the knots in the fabric of space-time, having boundaries and internal movements that are distinct, from the surrounding external Normal GraviTime.

    The Force of Gravity, presents itself as ripples or waves in the fabric of space GraviTime, and in its varying manifestations, is responsible for all Mass and Force.

7. There is never a circumstance of less than three entities:

    Nothingness,
    Uncertainty,
    and the relationship between the two.

8. Uncertainty always exists.
   It is the MOST fundamental entity.
   There is NEVER anything less than Uncertainty.
   Uncertainty gave birth to all Existence.
   Uncertainty trumps all else.

   Uncertainty can never be reduced, overcome, or fully known, not even by God. For even God can not be sure, that something else does not exist, that is just beyond the realm and reach of His knowledge.

9. If it moves, it has Mass.

   Since the fabric of space-time itself can bend, it can move, and has mass.

   Any mass that moves, simultaneously causes gravitational waves to be produced in its wake, which are imprinted on the fabric of space GraviTime itself. These waves travel outward at speeds up to those just below the speed of light. They are soon randomized along with the other randomized waves of the GraviTime vacuum. This randomized "quantum space foam," represents the small amount of heat that is inside space itself—the Cosmological Constant—which has a temperature that is just above absolute zero.

   All objects, by their very existence, contain definite boundaries that distinctly separate their internal motions and movements, from the external motions and movements of the surrounding space GraviTime, which lie outside their boundary.

   Such objects, contain a great number of individual particle movements inside their boundary. Because these movements are spherical, as are space Gravit themselves, the totalling of all internal object movements, along with the resulting gravitational waves they produce, create an average tendency for the movements to travel inward, toward the center of the object, and stay contained within (or adjacent to) the object.

   The result, is a "soaking up" of the surrounding space GraviTime, even if only for a moment, which creates the appearance of a bending of the space-time around the object, but which is in fact, its inward movement, at speeds up to those that are just below the speed of light. This inward movement creates a pressure, from the less active and more densely packed surrounding space GraviTime, and presents itself as the Force of Gravity.

   The gathering-up of the fabric of surrounding space GraviTime, from outside the object's boundary, is part of the ongoing recycling of surrounding space from the object's exterior, to places adjacent to or inside its boundary, which eventually return to its exterior, and move away from the object. This return, is accomplished through the outward-travelling radiation of electromagnetic energy, or by its sub-light ejection through other gravitational wave movements. This recycling occurs, for as long as the object exists separately and distinctly from the space GraviTime that surrounds it.

10. Force, is merely a quantity of movement in the fabric of space-time itself.

    All forces are differing manifestations of the more fundamental
    Force of Mass—space-time gravitational waves.

    The Electromagnetic Force, presents itself as a specific type of gravitational wave. Its corresponding matter particle—the electron—is its originator or result.

    An electron, is constructed of a knot of space GraviTime having **_uniform_** internal movement, the change of which creates an electromagnetic wave.

    Any change in the movement of an electron, causes its knot of GraviTime to unwind and punch through the space-time fabric, which initiates a self-repeating, chain-reaction gravitational wave. The motion of this wave, is contained within a narrow boundary of cohered movement, and may have no affect on the surrounding space GraviTime during its journey over potentially vast distances. The wave travels only at the speed of light, and its kinetic energy is not easily dissipated into the randomized waves of surrounding space-time, as it travels.

    By contrast, a generic gravitational wave (or GraviTime Wave), is initiated by the individual or bulk movement of any mass, which collides with an adjacent and equivalent mass of the surrounding Gravit and Tonne particles of space. This creates a ripple or wave in the fabric of space-time, which can quickly discohere into a wide boundary of movement. The wave can travel at any sub-light speed. Its kinetic energy can be more easily dissipated as it travels outward, away from the initiating source, eventually joining with and affecting, the whole remaining vast expanse of the randomly fluctuating fabric of space.

—Stony Brook
January 8, 2011

The author works as a consultant, and lives in Middletown, Connecticut 06457 USA; and Paris, Maine 04271 USA. The author was graduated from Bryant University in Smithfield, RI, 02917 USA. The author was graduated from Bryant University in 1989.

www.ingramcontent.com/pod-product-compliance
Lightning Source LLC
Chambersburg PA
CBHW040545220526
45473CB00016B/3027